The ESSENTIALS of

ADVANCED
CALCULUS II

**Staff of Research and Education Association,
Dr. M. Fogiel, Director**

This book is a continuation of *"THE
ESSENTIALS OF ADVANCED CALCULUS
I"* and begins with Chapter 6. It covers
the usual course outline of Advanced
Calculus II. Earlier/basic topics are
covered in *"THE ESSENTIALS OF
ADVANCED CALCULUS I".*

Research and Education Association
61 Ethel Road West
Piscataway, New Jersey 08854

THE ESSENTIALS OF ADVANCED CALCULUS II

Printed in the United States of America

Library of Congress Catalog Card Number 87-61819

International Standard Book Number 0-87891-568-0

WHAT "THE ESSENTIALS" WILL DO FOR YOU

This book is a review and study guide. It is comprehensive and it is concise.

It helps in preparing for exams, in doing homework, and remains a handy reference source at all times.

It condenses the vast amount of detail characteristic of the subject matter and summarizes the **essentials** of the field.

It will thus save hours of study and preparation time.

The book provides quick access to the important facts, principles, theorems, concepts, and equations of the field.

Materials needed for exams, can be reviewed in summary form — eliminating the need to read and re-read many pages of textbook and class notes. The summaries will even tend to bring detail to mind that had been previously read or noted.

This "ESSENTIALS" book has been carefully prepared by educators and professionals and was subsequently reviewed by another group of editors to assure accuracy and maximum usefulness.

Dr. Max Fogiel
Program Director

CONTENTS

This book is a continuation of *"THE ESSENTIALS OF ADVANCED CALCULUS I"* and begins with Chapter 6. It covers the usual course outline of Advanced Calculus II. Earlier/basic topics are covered in *"THE ESSENTIALS OF ADVANCED CALCULUS I"*.

Chapter No   **Page No.**

CHAPTER 6

MULTIPLE INTEGRALS

6.1 DOUBLE INTEGRALS

Let $f(x,y)$ be a function defined in a closed region, R, in the x-y plane (see Fig. 6.1). Subdivide R into N regions R_i of areas $\Delta A_1, \Delta A_2, \ldots, \Delta A_N$. Take M_i the biggest value of f in R_i, m_i the smallest value of f in R_i.

Let $\|A\|$ be $\max_{1 \le i \le N} (\Delta A_i)$.

Then if
$$\lim_{\substack{N \to \infty \\ \|A\| \to 0}} \sum_{i=1}^{N} M_i A_i = \lim_{\substack{N \to \infty \\ \|A\| \to 0}} \sum_{i=1}^{N} m_i A_i = I,$$

then f is called double integrable, and the value I is called the double integral of f. We write for I,

$$I = \iint_R f(x,y)dA = \iint_R f(x,y)dx\,dy. \qquad (6.1)$$

Properties of the Double Integral

a) $\displaystyle \iint_R cf(x,y)dA = c\iint_R f(x,y)dA.$

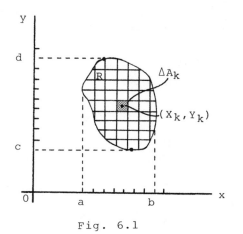

Fig. 6.1

(1)

b) $$\iint\limits_{R} [f(x,y) + g(x,y)]dA = \iint\limits_{R} f(x,y)dA + \iint\limits_{R} g(x,y)dA.$$

c) If $f(x,y)$ is integrable on a region R, it is integrable on any sub-region, $R_1, R_2, \ldots R_n$, of R, and if the R_i's do not overlap on a set of measure greater than zero and they cover R,

$$\iint\limits_{R} f(x,y)dA = \iint\limits_{R_1} f(x,y)dA + \iint\limits_{R_2} f(x,y)dA$$

$$+ \ldots + \iint\limits_{R_n} f(x,y)dA.$$

d) If $f(x,y)$ and $g(x,y)$ are integrable on R and if $f(x,y) \geq g(x,y)$ at each (x,y), then

$$\iint\limits_{R} f(x,y)dA \geq \iint\limits_{R} g(x,y)dA.$$

e) If $f(x,y)$ and $g(x,y)$ are integrable on R, and $g(x,y) \geq 0$, then there is a number μ

$$\inf_{R} f(x,y) \leq \mu \leq \sup_{R} f(x,y), \text{ such that}$$

$$\iint\limits_{R} f(x,y)g(x,y)dA = \mu \cdot \iint\limits_{R} g(x,y)dA.$$

63

f) If $f(x,y)$ is a continuous function on R, it is integrable on R.

g) If $f(x,y)$ is integrable on R, then so are $f(x,y)^+$, $f(x,y)^-$, and $|f(x,y)|$, where

$$f(x,y)^+ = \begin{cases} f(x,y) & f(x,y) \geq 0 \\ 0 & f(x,y) < 0 \end{cases}$$

$$f(x,y)^- = \begin{cases} 0 & f(x,y) \geq 0 \\ -f(x,y) & f(x,y) < 0 \end{cases}$$

$$|f(x,y)| = f(x,y)^+ + f(x,y)^-$$

$$f(x,y) = f(x,y)^+ - f(x,y)^-.$$

6.2 TRIPLE INTEGRALS

The concepts of the double integral are easily extended to closed regions in three-dimensions. Let $f(x,y,z)$ be a function defined in a closed region R, in the x-y-z plane (see Fig. 6.2). Subdivide the region into n regions, $R_1, \ldots, R_n$ of volume $\Delta V_1, \Delta V_2, \ldots, \Delta V_n$.

Take M_i the biggest value of f in R_i, m_i the smallest value of f in R_i.

Let $\|V\|$ be $\max_{1 \leq i \leq n} (V_i)$.

Then if

$$\lim_{\substack{n \to \infty \\ \|V\| \to 0}} \sum_{i=1}^{n} M_i V_i = \lim_{\substack{n \to \infty \\ \|V\| \to 0}} \sum_{i=1}^{n} m_i V_i = I,$$

then f is called triple integrable, and the value I is called the triple integral of f. For I, write

$$I = \iiint\limits_R f(x,y,z)dV = \iiint\limits_R f(x,y,z)dx \; dy \; dz. \qquad (6.2)$$

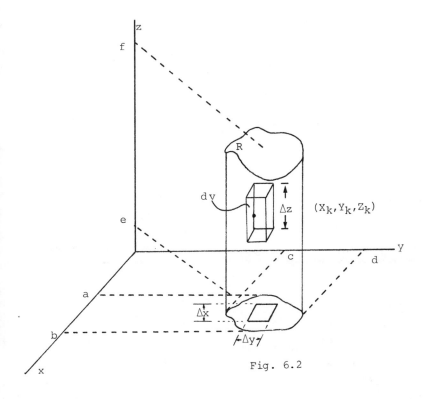

Fig. 6.2

The triple integral has all the properties of the double integral.

6.3 ITERATED INTEGRALS

Let $f(x,y)$ be a function defined on $\mathbb{R}$ in $\mathbb{R}^n$, and suppose for each y fixed, $f(x,y)$ is an integrable function of x. If $\int f(x,y)dx$ is an integrable function of y, its integral, $\int \left[\int f(x,y)dx \right] dy$, is called the iterated integral of $f(x,y)$.

In the continuous case, at least, this iterated integral is equal to the double integral. The above theorem (Fubini's theorem) indicates how a double integral can be evaluated.

If $f(x,y)$ is defined on the region

$R: \{ a \leq x \leq b, c \leq y \leq d \}$ (see Fig. 1), then

$$\iint_R f(x,y)dA = \iint_R f(x,y)dxdy = \int_c^d \left[\int_a^b f(x,y)dx \right] dy. \quad (6.3)$$

If $f(x,y,z)$ is defined on the region

$R: \{ a \leq x \leq b, c \leq y \leq d, e \leq z \leq f \}$, then

$$\iiint_R f(x,y,z)dV = \int_a^b dx \int_b^c dy \int_e^f f(x,y,z)dz. \quad (6.4)$$

More generally,

$$\int_R f(x_1,x_2,\ldots,x_n)dx_1\ldots dx_n$$
$$= \int_a^b \left[\int_{a_{n-1}(x_n)}^{b_{n-1}(x_n)} \ldots \left[\int_{a_1(x_2,\ldots,x_n)}^{b_1(x_2,\ldots,x_n)} f(x_1,\ldots,x_n)dx_1 \right] dx_2 \right] \ldots d_{x_{n-1}} dx_n, \quad (6.5)$$

The order in which integration occurs (i.e. first by dx_3, then by dx_1, then by dx_6, etc.) does not change the value of the integral when $f(x_1,\ldots,x_n)$ is continuous. It may be easier in some cases to write down the limits of integration for integrating in a different order.

6.4 CHANGE OF VARIABLES IN INTEGRALS

If (u_1,u_2) are curvilinear coordinates of points in a plane, the equations $x \equiv x(u_1,u_2)$ and $y = y(u_1,u_2)$ are a transformation of coordinates from the u_1, u_2 plane to the x,y plane. In such a case the region R of the xy plane is mapped by inverse functions into a region R* of the u_1, u_2 plane. Then the change of variable formula is

$$\iint_R f(x,y)dx\ dy = \iint_{R*} F(u_1,u_2)\left|\frac{\partial(x,y)}{\partial(u_1,u_2)}\right|\ du_1\ du_2 \qquad (6.6)$$

where $F(u_1,u_2) = f[x(u_1,u_2),y(u_1,u_2)]$ and

$$\frac{\partial(x,y)}{\partial(u_1,u_2)} = \begin{vmatrix} \dfrac{\partial x}{\partial u_1} & \dfrac{\partial x}{\partial u_2} \\ \\ \dfrac{\partial y}{\partial u_1} & \dfrac{\partial y}{\partial u_2} \end{vmatrix}. \qquad (6.7)$$

Similarly, if (u_1,u_2,u_3) are curvilinear coordinates in three dimensions, then

$$(6.8)$$

$$\iiint_R f(x,y,z)dx\ dy\ dz \qquad \iiint_{R*} F(u_1,u_2,u_3)\left|\frac{\partial(x,y,z)}{\partial(u_1,u_2,u_3)}\right|du_1\ du_2\ du_3$$

where

$$F(u_1,u_2,u_3) = f[x(u_1,u_2,u_3),y(u_1,u_2,u_3),z(u_1,u_2,u_3)]\ \text{and}$$

$$\frac{\partial(x,y,z)}{\partial(u_1,u_2,u_3)} = \begin{vmatrix} \dfrac{\partial x}{\partial u_1} & \dfrac{\partial y}{\partial u_1} & \dfrac{\partial z}{\partial u_1} \\ \\ \dfrac{\partial x}{\partial u_2} & \dfrac{\partial y}{\partial u_2} & \dfrac{\partial z}{\partial u_2} \\ \\ \dfrac{\partial x}{\partial u_3} & \dfrac{\partial y}{\partial u_3} & \dfrac{\partial z}{\partial u_3} \end{vmatrix}. \qquad (6.9)$$

CHAPTER 7

LINE INTEGRALS, SURFACE INTEGRALS, AND INTEGRAL THEOREMS

7.1 LINE INTEGRALS

A line integral is a kind of integral of a vector valued function which is defined along a curve C. In two dimensions, let $\vec{s}(t)$ be a vector valued function of the one real variable t, such that $\vec{s}(t)$ travels along the entire length of C, and $\vec{s}(0)$ = initial point of C, $\vec{s}(1)$ = final point of C. Divide the interval $[0,1]$ into n subintervals, $[a_0 = 0, a_1], [a_1, a_2], \ldots, [a_{n-1}, a_n = 1]$. Put $P_0, P_1, \ldots, P_n = \vec{s}(a_0)$ $\vec{s}(a_1), \ldots, \vec{s}(a_n)$, respectively. Put $Q_1, \ldots, Q_n = \vec{s}(c_1), \vec{s}(c_2), \ldots, \vec{s}(c_n)$ where C_i is in $[a_{i-1}, a_i]$.

Take a larger and larger number of subintervals, letting the length of each go to 0. Then

$$\lim_{n \to \infty} \sum_{k=1}^{n} \vec{f}(\vec{s}(Q_k)) \cdot (\vec{s}(P_k) - \vec{s}(P_{k-1})) = \int_C \vec{f} \cdot d\vec{s}$$

(7.1)

$$= \int_0^1 \left(\vec{f} \cdot \frac{d\vec{s}}{dt} \right) dt.$$

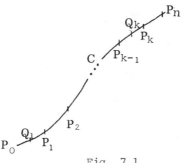

Fig. 7.1

The line integral exists if this limit does not depend on which C_i's we choose. Any function $\vec{s}$ which takes us along C in this way gives us the same value for $\int_C \vec{f} \cdot d\vec{s}$.

In three dimensions, the definition is the same, except $\vec{f}$ is a vector valued function of three variables, and s is a vector valued function of 1 real variable, i.e.

$$\vec{f}(x,y,z) = f_1(x,y,z)i + f_2(x,y,z)j + f_3(x,y,z)k$$
$$\vec{s}(t) = s_1(t)i + s_2(t)j + s_3(t)k. \tag{7.2}$$

A) Alternate form of the Line Integral

From equations (7.2) we can write

$$\int_C \vec{f} \cdot d\vec{s} = \int_C f_1 ds_1 + f_2 ds_2 + f_3 ds_3, \tag{7.3}$$

in which the variables and functions are often named differently, so that

$$\int_C \vec{f} \cdot d\vec{s} = \int_C P dx + Q dy + R dz \quad \text{(3 dimensions)} \tag{7.4}$$

$$\int_C \vec{f} \cdot d\vec{s} = \int_C P dx + Q dy \quad \text{(2 dimensions)}$$

This is an easy form to evaluate.

If $\vec{f}$ is a force acting on a mass, then

$$\int_C \vec{f}(x,y,z) \cdot d\vec{s}$$

69

is the work used to move the body along the entire length of C.

B) Evaluation of Line Integrals

The properties used in order to solve line integrals are the same as those of ordinary integrals. For example:

a) $\int_C f_1(x,y)dx + f_2(x,y)dy = \int_C f_1(x,y)dx + \int_C f_2(x,y)dy$

b) $\int_{P_0}^{P_n} f_1(x,y)dx + f_2(x,y)dy = \int_{P_0}^{P_1} f_1(x,y)dx + f_2(x,y)dy$

$$+ \int_{P_1}^{P_2} f_1(x,y)dx + f_2(x,y)dy$$

$$+ \ldots + \int_{P_{n-1}}^{P_n} f_1(x,y)dx + f_2(x,y)dy$$

c) $\int_{P_0}^{P_n} f_1(x,y)dx + f_2(x,y)dy = - \int_{P_n}^{P_0} f_1(x,y)dx + f_2(x,y)dy$

Line integrals, can be solved by the method of parametrization, which reduces the line integral to an ordinary integral. For example, if $x = f(t)$ and $y = g(t)$ the line integral $\int_C f_1(x,y)dx + f_2(x,y)dy$ becomes

$$\int_{t_1}^{t_2} f_1\{f(t),g(t)\}f'(t)dt + f_2\{f(t),g(t)\}g'(t)dt \qquad (7.5)$$

Another method uses the fact that certain line integrals are independent of the curve C. These line integrals have the same value between two endpoints in space along any curve. In this case form (7.4) of the line integral is an integral of an exact differential, and we can find a potential function, which gives the value of the line integral from the initial point p_0 to the terminal point p_1.

Another method to solve line integrals is by Green's Theorem which transforms the line integral around some closed curve to a double integral on the region bounded by this curve.

7.2 INDEPENDENCE OF PATH

Let $f_1(x,y)$ and $f_2(x,y)$ have continuous partial derivatives in a domain D and let D be simply connected. If

$$\frac{\partial f_2(x,y)}{\partial x} = \frac{\partial f_1(x,y)}{\partial y}$$

in D, then

$$\int_C f_1(x,y)dx + f_2(x,y)dy$$

is independent of the path in D.

In a similar way, the line integral

$$\int_C f_1(x,y,z)dx + \int_C f_2(x,y,z)dy$$

$$+ \int_C f_3(x,y,z)dz \tag{7.6}$$

is independent of the path in D if

$$\frac{\partial f_1(x,y,z)}{\partial y} = \frac{\partial f_2(x,y,z)}{\partial x} \tag{7.7}$$

$$\frac{\partial f_3(x,y,z)}{\partial x} = \frac{\partial f_1(x,y,z)}{\partial z} \tag{7.8}$$

$$\frac{\partial f_2(x,y,z)}{\partial z} = \frac{\partial f_3(x,y,z)}{\partial y}. \tag{7.9}$$

In vector form, since

$$\nabla \times \vec{f}(x,y,z) = \left[\left(\frac{\partial f_3(x,y,z)}{\partial y} - \frac{\partial f_2(x,y,z)}{\partial z} \right) i \right.$$

$$+ \left(\frac{\partial f_1(x,y,z)}{\partial z} - \frac{\partial f_3(x,y,z)}{\partial x} \right) j$$

$$\left. + \left(\frac{\partial f_2(x,y,z)}{\partial x} - \frac{\partial f_1(x,y,z)}{\partial y} \right) k \right]$$

the condition for $f(x,y,z)$ to be independent of the path in D is

$$\nabla \times \vec{f}(x,y,z) = \vec{0}, \tag{7.10}$$

where $\vec{f}(x,y,z) = f_1(x,y,z)i + f_2(x,y,z)j + f_3(x,y,z)k$

A sufficient condition for

$$\int_C f_1(x,y,z)dx + f_2(x,y,z)dy + f_3(x,y,z)dz$$

be independent of the path is that

$$f_1(x,y,z)dx + f_2(x,y,z)dy + f_3(x,y,z)dz \tag{7.11}$$

must be an exact differential.

This is an exact differential if there exists a function $\phi(x,y,z)$ such that

$$f_1(x,y,z)i + f_2(x,y,z)j + f_3(x,y,z)k = \nabla \phi(x,y,z) \tag{7.12}$$

In this case if the points P_0 and P_n have coordinates (x_0,y_0,z_0) and (x_n,y_n,z_n) respectively, the value of the line integral is

$$\int_C \vec{f} \cdot d\vec{s} = \int_{(x_0,y_0,z_0)}^{(x_n,y_n,z_n)} d\phi = \phi(x_n,y_n,z_n) - \phi(x_0,y_0,z_0). \tag{7.13}$$

In vector form,

$$\int_{(x_0,y_0,z_0)}^{(x_n,y_n,z_n)} f(x,y,z)dr = \phi(x_n,y_n,z_n) - \phi(x_0,y_0,z_0),$$

where $\qquad$ dr = dxi + dyj + dzk. $\qquad$ (7.14)

7.3 GREEN'S THEOREM

Let R be a region in $\mathbb{R}^2$ and let C be the curve bounding R (see Fig. 7.2). C is oriented in such a way that R is always to the left of C while travelling in the positive direction. Let $f_1(x,y)$, $f_2(x,y)$ be real-valued continuously differentiable functions defined on R. Then Green's Theorem in the plane states that,

$$\oint_C f_1(x,y)dx + f_2(x,y)dy = \iint_R \left[\frac{\partial f_2(x,y)}{\partial x} - \frac{\partial f_1(x,y)}{\partial y} \right] dx\, dy.$$

(7.15)

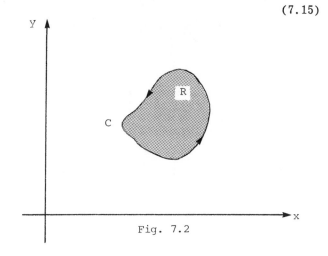

Fig. 7.2

7.4 SURFACE INTEGRALS

A surface integral is a kind of integral of a real valued function which is defined on a surface. Let S be a surface, and let $\phi(Q)$ be a real valued function which is

defined at all points on S. Let the elements ΔA_1, ΔA_2,....
ΔA_n divide the surface S into n surface elements. Let Q_k
be some point in the element ΔA_k (Fig. 7.3), k =
1, 2, 3,...,n. The limit of the sum

$$\sum_{k=1}^{n} \phi(Q_k) \Delta A_k$$

as, n → ∞ in such a way as ΔA_k → 0, is called the surface
integral of ϕ over S, or

$$\iint_S \phi \, dA = \lim_{n \to \infty} \sum_{k=1}^{n} \phi(Q_k) \Delta A_k \qquad (7.16)$$

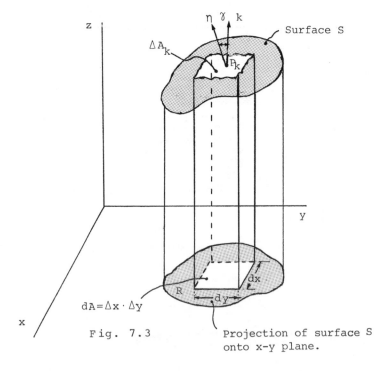

Fig. 7.3

Projection of surface S
onto x-y plane.

The surface integral exists if this limit does not depend on
which Q_k we pick. If the function $\phi(Q)$ is expressed

74

as f(x,y,z) the above integral may be written as

$$\iint_S f(x,y,z)dA \qquad (7.17)$$

Take $\vec{u}(x,y)$ $(0 \leq x \leq 1, 0 \leq y \leq 1)$ a function which parameterizes S, i.e. it is always on S and it covers S completely. Then

$$\iint_S f(x,y,z)dA = \iint_R f(\vec{u}(x,y)) \left\| \frac{d\vec{u}}{dx} \times \frac{d\vec{u}}{dy} \right\| dx\,dy \qquad (7.18)$$

where R is the square $(0 \leq x \leq 1, 0 \leq y \leq 1)$.

In applications, the following case often comes up: The surface S is defined as a level curve of g(x,y,z), say g(x,y,z) = 0. We are given a vector valued function $\vec{F}(x,y,z)$ and are asked to find a surface integral through S. This means find

$$\iint_S f(x,y,z)dA \quad \text{where} \quad f(x,y,z) = \vec{F}(x,y,z) \cdot \vec{n}(x,y,z),$$

$\vec{n}(x,y,z)$ unit normal to S, $\vec{n} = \pm \dfrac{\nabla g}{|\nabla g|}$.

The evaluation of a surface integral is accomplished by reducing it to a double integral. This is done by various methods depending upon the representation of the surface. Surface integrals can also be evaluated by the Divergence Theorem, Stoke's Theorem, and the change of variable formula.

Surface integrals are useful in formulation of physical concepts such as finding area, centers of gravity, moments of inertia of curved laminae, and other important physical quantities.

7.5 THE DIVERGENCE THEOREM

Let a region R of volume V be bounded by a closed surface S. Let $f_1(x,y,z)$, $f_2(x,y,z)$, $f_3(x,y,z)$ be continuous and have partial derivatives in the region including the surface S and its boundary. Then

$$\iiint\limits_{R} \left(\frac{\partial f_1}{\partial x} + \frac{\partial f_2}{\partial y} + \frac{\partial f_3}{\partial z} \right) dv = \oiint\limits_{S} (f_1 \cos\alpha + f_2 \cos\beta + f_3 \cos\gamma) ds$$

$$(7.19)$$

where α, β, γ the angles which the normal vector $\vec{n}(\vec{n} = \cos\alpha i + \cos\beta j + \cos\gamma k)$ makes with the positive x, y and z axes respectively.

In vector form, with $\vec{F} = f_1 i + f_2 j + f_3 k$ equation (7.19) can be written as:

$$\iiint\limits_{R} \nabla \cdot \vec{F} \, dv = \oiint\limits_{S} \vec{F} \cdot \vec{n} \, ds \qquad (7.20)$$

7.6 STOKE'S THEOREM

Let S be a smooth surface bounded by a closed non-interesting curve C. Let S be oriented and assume that the boundary curve is oriented so that S lies to the left of C. Let $f_1(x,y,z)$, $f_2(x,y,z)$ and $f_3(x,y,z)$ be single-valued continuous functions which have continuous first partial derivatives in a region of space including S and its boundary. Then,

$$\int\limits_{C} f_1 dx + f_2 dy + f_3 dz = \iint\limits_{S} \left[\left(\frac{\partial f_3}{\partial y} - \frac{\partial f_2}{\partial z} \right) \cos\alpha \right.$$

$$\left. + \left(\frac{\partial f_1}{\partial z} - \frac{\partial f_3}{\partial x} \right) \cos\beta + \left(\frac{\partial f_2}{\partial x} - \frac{\partial f_1}{\partial y} \right) \cos\gamma \right] ds$$

$$(7.21)$$

In vector form, with $\vec{F} = f_1 i + f_2 j + f_3 k$ and $\vec{n} = \cos\alpha i + \cos\beta j + \cos\gamma k$ (see Fig. 7.4), the Stokes' equation may be written as

$$\int\limits_{C} \vec{F} \cdot d\vec{r} = \oiint\limits_{S} (\nabla \times \vec{F}) \cdot \vec{n} \, ds \qquad (7.22)$$

where $\qquad$ $d\vec{r} = dxi + dyj + dzk$

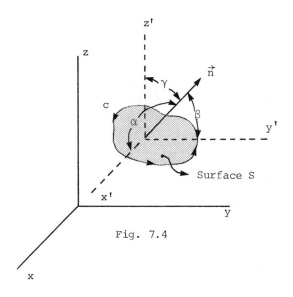

Fig. 7.4

CHAPTER 8

INFINITE SERIES

8.1 DEFINITION OF INFINITE SERIES

If $a_0, a_1, a_2, \ldots$ is an infinite sequence then an infinite series is the sum

$$\sum_{n=0}^{\infty} a_n = a_0 + a_1 + a_2 + \ldots \qquad (8.1)$$

The numbers $a_0, a_1, a_2, \ldots$ are the terms of the series, where a_n is defined as the general term.

For an infinite series, $\sum_{n=0}^{\infty} a_n$, to be of practical use, it must be convergent.

If the partial sum, $S_n = a_0 + a_1 + \ldots + a_n$ is given and $\lim_{n \to \infty} S_n = L$, then the series $\sum_{n=0}^{\infty} a_n$ converges to L. A series that is not convergent is called divergent.

Two kinds of series which are often used are:

A) Geometric series: $\boxed{\sum_{n=1}^{\infty} ar^{n-1} = a + ar + ar^2 + \ldots}$ where

a and r are constants. This series converges to $s = \dfrac{a}{1-r}$ if $|r| < 1$ and diverges if $|r| \geq 1$. The sum of the first n terms is given by $S_n = \dfrac{a(1 - r^n)}{1 - r}$.

B) The k series $\sum\limits_{n=0}^{\infty} \dfrac{1}{n^k} = \dfrac{1}{1^k} + \dfrac{1}{2^k} + \dfrac{1}{3^k} + \ldots$, where k is a constant. This series is convergent for $k > 1$ and divergent for $k \leq 1$. For $k = 1$, the series becomes $1 + \dfrac{1}{2} + \dfrac{1}{3} + \dfrac{1}{4} + \ldots$, which diverges and is called the harmonic series.

8.2 CONVERGENCE AND DIVERGENCE OF INFINITE SERIES

A) Theorems

a) if the series, $\sum\limits_{n=0}^{\infty} a_n$, converges then $\lim\limits_{n \to \infty} a_n = 0$.
However, if $\lim\limits_{n \to \infty} a_n = 0$, then the series $\sum\limits_{n=0}^{\infty} a_n$ may or may not converge.

b) The series $\sum\limits_{n=0}^{\infty} a_n$ converges if the sequence of partial sums S_n is increasing and bounded.

c) The convergence or divergence of a series is not affected by the removal of a finite number of terms, or the multiplication of each term by a constant number different from zero.

d) If the series $\sum\limits_{n=0}^{\infty} a_n$ is convergent and so is $\sum\limits_{n=0}^{\infty} |a_n|$, the $\sum\limits_{n=0}^{\infty} a_n$ is called absolutely convergent. If the series converges, but the series of absolute values diverges, then the series is called conditionally convergent. If the series of absolute values converges then so does the series.

e) A series $\sum\limits_{n=0}^{\infty} a_n$ converges if for all $\varepsilon > 0$ there is a positive number N for which $|a_{n+1} + a_{n+2} + \ldots + a_m| < \varepsilon$ for all m and n such that $m > n > N$.

f) The sum, difference, and product of two absolutely convergent series is absolutely convergent.

B) Comparison Test

    a) convergence

        Suppose that Σa_n is a convergent series with $a_n > 0$ for all n, then a series Σb_n also converges if $a_n \geq b_n \geq 0$ for all $n > N$, for some N.

    b) divergence

        if Σa_n diverges and $a_n \leq b_n$ for all $n > N$, then Σb_n also diverges.

C) Quotient Test

    If Σa_n and Σb_n are two series of positive terms and $0 <$

$$\lim_{n \to \infty} \frac{a_n}{b_n} = L < \infty,$$ then either both series are convergent, or both are divergent. If $L = 0$ and Σb_n converges, so does Σa_n.

    If $L = \infty$ and Σb_n diverges, so does Σa_n.

D) Integral test

    Let $f(x)$ be a function which is positive, continuous and non-increasing as x increases for all values of $x \geq N$, where N is some fixed positive integer. Then the series $\Sigma a_n = \Sigma f(n)$ converges or diverges according to the integral

$$\lim_{R \to \infty} \int_N^R f(x)dx \qquad (*)$$

If $(*)$ converges or diverges, then the series Σa_n converges or diverges correspondingly.

E) Ratio Test

If the series Σa_n has positive terms, and if $\lim\limits_{n \to \infty} \left| \dfrac{a_{n+1}}{a_n} \right|$ = L, then the series Σa_n converges absolutely if L < 1, and Σa_n diverges if L > 1. L = 1 the test is inconclusive.

F) Raabe's Test

If Σa_n is a series with positive terms, and if $\lim\limits_{n \to \infty} n \left(1 - \dfrac{a_{n+1}}{a_n} \right)$ = L, then the series Σa_n converges absolutely if L > 1 and diverges or converges conditionally for L < 1. If L = 1, the test is inconclusive.

G) Gauss' Test

If the ratio $\left| \dfrac{a_{n+1}}{a_n} \right|$ can be expressed in the form $1 - \dfrac{L}{n} + \dfrac{C_n}{n^q}$ where $|C_n| < p$ for all n > N and q > 1, then the series Σa_n is absolutely convergent if L > 1, and diverges or converges conditionally if $L \leq 1$.

H) Cauchy's Root Test

If the $\lim\limits_{n \to \infty} |a_n|^{\frac{1}{n}}$ = L the series Σa_n converges absolutely if L < 1 and diverges if L > 1. For L = 1 the test is inconclusive.

I) Alternating Test

If the terms of a series are alternately positive and negative, $|a_{n+1}| \leq |a_n|$ and $\lim\limits_{n \to \infty} a_n = 0$, then the series Σa_n is convergent.

8.3 SERIES OF FUNCTIONS, UNIFORM CONVERGENCE

A function can be defined as a sum of an infinite series, or

$$f(x) = \sum_{n=1}^{\infty} a_n(x),$$

where the terms of the sequence of functions are defined in a closed interval $[a,b]$.

If the limit of the sequence of partial sums exists for all $x \in [a,b]$ then the series is said to be pointwise convergent in $[a,b]$.

The sum $S_n(x)$ is the nth partial sum. By recalling the definition of the limit, it follows that $\sum_{n=1}^{\infty} a_n(x)$ converges pointwise to $f(x)$ in $[a,b]$ if for all $\varepsilon > 0$ and each $x \in [a,b]$ there exists an $N > 0$ such that $|S_n(x) - f(x)| < \varepsilon$ for all $n > N$. According to the above definition the number N depends on ε and x. If the number N depends on ε and not on x, the series is called uniformly convergent on $[a,b]$.

8.4 COMPARISON TESTS FOR UNIFORM CONVERGENCE

A) Weierstrass M Test

A sufficient but not a necessary condition for a series $\sum_{n=1}^{\infty} a_n(x)$ to be uniformly convergent in a defined interval is the relation

$$|S_n(x)| \leq M_n,$$

where $S_n(x) = a_1(x) + a_2(x) + \ldots + a_n(x)$, and $M_n = M_1 + M_2 + M_3 + \ldots$ is a convergent series of positive constants.

B) Dirichlet's Test

The series $\sum_{n=1}^{\infty} b_n a_n(x)$ is uniformly convergent in $[a,b]$ if

a) there exists a constant P such that $|S_n(x)| < P$ for all $n > N$ and

b) the sequence $\{b_n\}$ is a monotonic decreasing sequence of positive constants having limit zero.

8.5 POWER SERIES, CONVERGENCE

One of the most important series is the power series. A power series in powers of $(x - c)$ is a series of the form

$$\sum_{n=0}^{\infty} a_n(x - c)^n = a_0 + a_1(x - c) + a_2(x - c)^2 + \ldots \qquad (8.2)$$

Letting $c = 0$, the power series takes the form

$$\sum_{n=0}^{\infty} a_n x^n = a_0 + a_1 x + a_2 x^2 + \ldots \qquad (8.3)$$

Of importance in the study of power series is the notion of radius of convergence, R, which can be often found by the ratio test. In general, a power series converges for $|x| <$ R, diverges for $|x| >$ R, and may or may not converge for $|x| =$ R. For R = 0, the series converges only for x = 0. If R = ∞ the series converges for all values of x.

The series $\sum_{n=0}^{\infty} a_n x^n$ is absolutely and uniformly convergent if $|x| <$ R.

For the two convergent power series $\sum_{n=0}^{\infty} a_n x^n$ and $\sum_{n=0}^{\infty} b_n x^n$, the following theorems are valid:

A) The series can be added or subtracted term by term.

B) These series can be multiplied to obtain the series $\sum_{n=0}^{\infty} c_n x^n$, where

$$c_n = \sum_{k=0}^{\infty} a_k b_{n-k} = a_0 b_n + a_1 b_{n-1} + \ldots + a_n b_0$$

C) The series $\sum\limits_{n=0}^{\infty} a_n x^n$ can be divided by the series $\sum\limits_{n=0}^{\infty} b_n x^n$ ($b_0 \neq 0$) to obtain the series $\sum\limits_{n=0}^{\infty} c_n x^n$, whose coefficients may be found by the process of division or by solving the system

$$b_0 c_0 = a_0$$

$$b_0 c_1 + b_1 c_0 = a_1$$
$$\vdots$$

for $\qquad c_0, c_1, \ldots, c_n$.

D) If $\sum\limits_{n=0}^{\infty} a_n x^n = \sum\limits_{n=0}^{\infty} b_n x^n$, then $a_n = b_n$ for all n.

8.6 THEOREM ON POWER SERIES

A) Abel's Theorem

If the series $\sum\limits_{n=0}^{\infty} a_n x^n$ converges at R, then it converges uniformly on the closed interval $0 \leq x \leq R$. The conclusion holds for $-R \leq x \leq 0$ if the series converges at $x = -R$.

B) Abel's Limit Theorem

If the series $\sum\limits_{n=0}^{\infty} a_n x^n$ converges at $x = x_0$, where x_0 may be an interior point or an endpoint of the interval of convergence, then

$$\lim_{x \to x_0} \left\{ \sum\limits_{n=0}^{\infty} a_n x^n \right\} = \sum\limits_{n=0}^{\infty} \lim_{x \to x_0} a_n x^n = \sum\limits_{n=0}^{\infty} a_n x_0^n$$

If x_0 is a left-hand endpoint it is proper to write $x \to x_0^+$ and for a right-hand endpoint $x \to x_0^-$.

C) If the series $\sum\limits_{n=0}^{\infty} a_n x^n$ has a radius of convergence $R >$

0, then the series converges uniformly on the closed interval $[-r,r]$, where $0 \leq r < R$.

D) If a function is defined as

$$f(x) = \sum_{n=0}^{\infty} a_n x^n,$$

then the integral of this function $\left[\int_{a}^{b} f(x)dx \right]$ is equal to the series obtained by integrating the original power series term by term, or

$$\int_{a}^{b} \left(\sum_{n=0}^{\infty} a_n x^n \right) dx = \sum_{n=0}^{\infty} \frac{a_n}{n+1} (b^{n+1} - a^{n+1})$$

If the function $f(x) = \sum_{n=0}^{\infty} a_n x^n$ has a radius of convergence $R > 0$, then it is differentiable term by term or,

$$f'(x) = \sum_{n=0}^{\infty} a_n \frac{d}{dx}(x^n) = \sum_{n=1}^{\infty} na_n x^{n-1}$$

for $|x| < R$.

8.7 TAYLOR AND MacLAURIN SERIES

If the function $f(x)$ is continuous in a closed interval $[a,b]$ and its derivatives $f'(x), f''(x), \ldots, f^n(x)$ and $f^{n+1}(x)$ exist, then the function $f(x)$ can be represented by a series expansion of $f(x)$ about the point $x = a$, or

$$\boxed{f(x) = \sum_{n=0}^{\infty} \frac{f^n(a)}{n!} (x - a)^n} \qquad (8.4)$$

The function may also be represented by $\sum_{k=0}^{\infty} \dfrac{f^k(a)}{K!} + R_{n+1}$ where R_{n+1} is the remainder.

Lagrange's form of the remainder is

$$R_{n+1} = \frac{f^{n+1}(\zeta)}{(n + 1)!} (x - a)^{n+1} \tag{8.5}$$

Cauchy's form is

$$R_{n+1} = \frac{f^{n+1}(\zeta)}{n!} (x - \zeta)^n (x - a)$$

$$x < \zeta < a \tag{8.6}$$

If $\lim\limits_{n\to\infty} R_{n+1} = 0$, then the series (8.4) is called the Taylor series.

If $a = 0$, equation (8.4) often is called the MacLaurin series expansion of $f(x)$.

CHAPTER 9

IMPROPER INTEGRALS

9.1 DEFINITION OF AN IMPROPER INTEGRAL

By definition, an integral is improper if at least one of the limits is infinite, or if the integrand has one or more points of discontinuity in the interval of integration, or if both of the above conditions exist.

Respectively the above statement defines improper integrals of the first, second and third (or mixed) kind.

There are many analogies between the theory of improper integrals and that of infinite series. That is, improper integrals can be termed convergent or divergent in the same manner as infinite series. Furthermore the notions of absolute convergence and conditional convergence exist for improper integrals. In addition, if an improper integral defines a function, it can be determined whether or not this integral is uniformly convergent on a given interval.

9.2 IMPROPER INTEGRALS OF THE FIRST KIND WITH NON-NEGATIVE TERMS

An improper integral of the first kind, is an integral of the form

$$\int_a^\infty f(x)dx \qquad (a \leq x \leq \infty) \qquad\qquad (9.1)$$

The value of this integral is defined as the limit

$$\lim_{b \to \infty} \int_a^b f(x)dx \qquad\qquad (9.2)$$

If the limit exists, then the integral (9.1) is said to be convergent; otherwise it is divergent.

Examples

A) The geometric or exponential integral $\int_a^b e^{-rx}dx$, where r is a constant.

Applying equation (9.2),

$$\lim_{b \to \infty} \int_a^b e^{-rx}dx = \lim_{b \to \infty} \left. \frac{-e^{-rx}}{r} \right|_a^b = \lim_{b \to \infty} \left(-\frac{e^{-rb}}{r} + \frac{e^{-ra}}{r} \right)$$

which exists and is equal to $\dfrac{e^{-ra}}{r}$ if $r > 0$ and does not exist if $r \leq 0$.

Therefore $\int_a^\infty e^{-rx}dx$ converges if $r > 0$ and diverges if $r \leq 0$.

B) The p integral of the first kind $\int_a^\infty \dfrac{dx}{x^p}$ where p is a constant and $a > 0$.

Applying equation (9.2),

$$\lim_{b \to \infty} \int_a^b \frac{dx}{x^p} = \lim_{b \to \infty} \left. \frac{x^{1-p}}{1-p} \right|_a^b = \lim_{b \to \infty} \left[\frac{b^{1-p}}{1-p} - \frac{a^{1-p}}{1-p} \right]$$

88

which exists and is equal to $\dfrac{-a^{1-p}}{1-p}$ for $p > 1$ and does not exist for $p \leq 1$.

Therefore $\displaystyle\int_a^\infty \dfrac{dx}{x^p}$ converges if $p > 1$ and diverges if $p \leq 1$.

9.3 CONVERGENCE AND DIVERGENCE FOR IMPROPER INTEGRALS OF THE FIRST KIND

A) Theorems

a) The integral, $\displaystyle\int_a^\infty f(x)dx$ with $f(x) > 0$ is convergent if and only if there is a constant number N such that

$$\int_a^t f(x)dx \leq N \quad \text{when } t > a.$$

b) The integral $\displaystyle\int_a^\infty f(x)dx$ is absolutely convergent, if

$\displaystyle\int_a^\infty |f(x)|dx$ is convergent.

If $\displaystyle\int_a^\infty |f(x)|dx$ converges, then $\displaystyle\int_a^\infty f(x)dx$ converges. If

$\displaystyle\int_a^\infty f(x)dx$ converges but $\displaystyle\int_a^\infty |f(x)|dx$ diverges, then the integral is called conditionally convergent.

c) The integral $\int_a^\infty g(x)f(x)dx$ is convergent if

   1) $\dfrac{dg(x)}{dx}$ is continuous, $\dfrac{dg(x)}{dx} \leq 0$, and $\lim\limits_{x \to \infty} g(x) = 0$

   2) $f(x)$ is continuous and $F(x) = \int_a^t f(x)dx$

   is bounded for all $t \geq a$.

d) If $g(x) = x^P$ and $\lim\limits_{x \to \infty} x^P f(x) = L$, then

$\int_a^\infty f(x)dx$ converges if $p > 1$ and L is finite, and

$\int_a^\infty f(x)dx$ diverges if $p \leq 1$ and $L \neq 0$.

B) Comparison Test

  a) Convergence

    Suppose that $\int_a^\infty g(x)dx$, $g(x) \geq 0$ for all $x \geq a$,

    converges, then the integral $\int_a^\infty f(x)dx$ also

    converges if $g(x) \geq f(x) \geq 0$ for all $x \geq a$.

  b) Divergence

    If $\int_a^\infty g(x)dx$, $g(x) \geq 0$ diverges for all $x \geq a$, then the integral $\displaystyle\int_a^\infty f(x)dx$ also diverges if $f(x) \geq g(x)$ for all $x \geq a$.

C) Quotient Test

Suppose $\displaystyle\int_a^\infty f(x)dx$ and $\displaystyle\int_b^\infty g(x)dx$ are integrals of the first kind with $f(x) \geq 0$ and $g(x) \geq 0$ respectively, then

a) If the limit $\displaystyle\lim_{x \to \infty} \frac{f(x)}{g(x)} = L$ exists (finite) and is not zero, then either both integrals are convergent or both are divergent.

b) If $L = 0$ and $\displaystyle\int_b^\infty g(x)dx$ is convergent, so is $\displaystyle\int_a^\infty f(x)dx$.

c) If $L = +\infty$ and $\displaystyle\int_b^\infty g(x)dx$ diverges, so does $\displaystyle\int_b^\infty f(x)dx$.

9.4 IMPROPER INTEGRALS OF THE SECOND KIND

An improper integral of the second kind exists if the function $f(x)$ is unbounded at one or more points of $a \leq x \leq b$. These points are termed singularities of $f(x)$.

If the function $f(x)$ has a singularity, x_0, $(a < x_0 < b)$, then the value of the improper integral is defined as

$$\lim_{\varepsilon_1 \to 0^+} \int_a^{x_0 - \varepsilon_1} f(x)dx + \lim_{\varepsilon_2 \to 0^+} \int_{x_0 + \varepsilon_2}^b f(x)dx \qquad (9.3)$$

If the $\displaystyle\lim_{\varepsilon_1 \to 0^+} \int_a^{x_0 - \varepsilon_1} f(x)dx$ exists, the integral is called the left convergent of the point x_0. Similarly if the

$\displaystyle\lim_{\varepsilon_1 \to 0} \int_{x_0 + \varepsilon_2}^b f(x)dx$ exists, the integral is called the right convergent of the point x_0.

Extensions of these definitions can be made in case $f(x)$ becomes unbounded at two or more points of the interval.

The most common textbook integrals of the second kind are:

$$\int_a^b \frac{dx}{(b-x)^p} \, , \qquad \int_a^b \frac{dx}{(x-a)^p} \qquad (9.4)$$

These integrals are convergent if $p < 1$, and divergent if $p \geq 1$. If $p \leq 0$ they are proper integrals with no singularities of the integrands.

9.5 CONVERGENCE AND DIVERGENCE TESTS FOR IMPROPER INTEGRALS OF THE SECOND KIND

A) Comparison Test

a) Convergence

Let $f(x)$ be bounded at $x = x_0$. Suppose that $\displaystyle\int_a^b g(x)dx$, $g(x) \geq 0$ for all $a \leq x \leq b$, converges,

then the integral $\displaystyle\int_a^b f(x)dx$ also converges if $0 \leq$

$f(x) \le g(x)$ for $a \le x \le b$.

b) Divergence

If $\displaystyle\int_a^b g(x)dx$, $g(x) \ge 0$ diverges for all $a \le x \le b$,

then the integral $\displaystyle\int_a^b f(x)dx$ also diverges if $f(x) \ge$

$g(x)$ for $a \le x \le b$.

B) Quotient Test

If $f(x) \ge 0$ and $g(x) \ge 0$ for $a \le x \le b$, then:

a) The integrals $\displaystyle\int_a^b f(x)dx$ and $\displaystyle\int_a^b g(x)dx$ both

converge or diverge if the $\displaystyle\lim_{x \to \infty} \frac{f(x)}{g(x)} = L$ exists and is not zero.

b) If $L = 0$ and $\displaystyle\int_a^b g(x)dx$ converges, then $\displaystyle\int_a^b f(x)dx$ converges.

c) If $L = \infty$ and $\displaystyle\int_a^b g(x)dx$ diverges, then

$\displaystyle\int_a^b f(x)dx$ diverges.

C) Absolute and Conditional Convergence

If $\displaystyle\int_a^b |f(x)|dx$ converges, then $\displaystyle\int_a^b f(x)dx$ is called absolutely convergent.

If $\displaystyle\int_a^b f(x)dx$ converges but $\displaystyle\int_a^b |f(x)|dx$ diverges, then $\displaystyle\int_a^b f(x)$ is called conditionally convergent.

9.6 IMPROPER MIXED INTEGRALS

Many improper integrals occuring in practice can be expressed in terms of sums of improper integrals of the first and second kinds. The convergence and divergence of these integrals is determined by using the theorem and tests applied to improper integrals of the first and second kinds.

9.7 UNIFORM CONVERGENCE FOR IMPROPER INTEGRALS

Suppose that the integral $\displaystyle F(t) = \int_a^\infty f(x,t)dx$ converges for each fixed t in the interval $A \leq x \leq B$. In addition let $\displaystyle S_R(t) = \int_a^R f(x,t)dx$. Then, by definition, the integral $S_R(t)$ converges uniformly, if for each $\varepsilon > 0$ there exists an N depending only on ε such that

$$\left| F(t) - S_r(t) \right| < \varepsilon \quad \text{for all } R > N \text{ and } A \leq t \leq B$$

An analogous definition can be given for improper integrals of the second kind for mixed integrals.

A) Weierstrass M Test for Uniform Convergence

The integral $\displaystyle F(t) = \int_a^\infty f(x,t)dx$ is uniformly and absolutely convergent in $A \leq t \leq B$ if

a) A function $M(x) \geq 0$ can be found such that $\left| f(x,t) \right| \leq M(x)$

b) $M(x)$ and $f(x,t)$ are continuous

c) $\displaystyle\int_a^\infty M(x)dx$ converges

B) Dirichlet's Test for Uniform Convergence

The integral $\displaystyle\int_a^\infty g(x,y)f(x,y)dx$ converges uniformly on $A \le y \le B$ if

a) $g(x,y)$ is continuous on $a \le x$, $A \le y \le B$

b) $\displaystyle\left| \int_a^R g(x,y)dx \right| < k$ (k = constant) for all $R \ge a$ and all y on $[A,B]$

c) $f(x,y)$ is a decreasing function of x for $x \ge a$ ($A \le y \le B$) and approaches zero uniformly in y as $x \to \infty$.

C) Theorems on Uniformly Convergent Integrals

a) If $f(x,t)$ is continuous on $c \le x$, $a \le t \le b$, and if

$$F(t) = \int_c^\infty f(x,t)dx$$

is uniformly convergent for $a \le t \le b$, then $F(t)$ is continuous on the interval $a \le t \le b$.

b) Under the hypothesis of theorem (a), it is true that

$$\int_a^b F(t)dt = \int_c^\infty \left\{ \int_a^b f(x,t)dt \right\} dx$$

c) If $f(x,t)$ is continuous on $c \le x$, $a \le t \le b$, and if

$$\int_c^\infty \frac{\partial f}{\partial t} dx$$

converges uniformly on $a \le t \le b$, then $F(t)$ has a derivative given by

$$F'(t) = \int_c^\infty \frac{\partial f}{\partial t} dx$$

9.8 LAPLACE TRANSFORMS

The Laplace transform of a function is defined by

$$f(s) = L\{F(x)\} = \int_0^\infty e^{-sx}F(x)dx \qquad (9.5)$$

Laplace transforms are used in solving differential equations and in some special topics in mathematics by using the following procedure:

(original problem) $\xrightarrow{\text{L}}$ (Transformed problem)

(solution of original problem) $\xleftarrow{\text{L}^{-1}}$ (Solution of transformed problem)

If two functions $f(x)$ and $g(x)$ have Laplace transforms $F(s)$ and $G(s)$ respectively, then

$$L\left[\int_0^x g(x-x')f(x')dx'\right] = G(s)F(s) \qquad (9.6)$$

This is known as the convolution theorem. The integral on the left is called the convolution of g and f, written $g(x) * f(x)$.

9.9 GAMMA AND BETA FUNCTIONS

A) The Gamma Function

The gamma function, $\Gamma(t)$, is defined for any positive real number t by

$$\Gamma(t) = \int_0^\infty x^{t-1}e^{-x}dx \qquad (9.7)$$

A simple relation between the values of the gamma

function at t and t + 1 is

$$\Gamma(t + 1) = t\,\Gamma(t) \tag{9.8}$$

From this formula for $t = 1, 2, 3, \ldots$

$$\Gamma(2) = 1 \cdot \Gamma(1) = 1$$
$$\Gamma(3) = 2 \cdot \Gamma(2) = 2 \cdot 1$$
$$\Gamma(4) = 3 \cdot \Gamma(3) = 3 \cdot 2 \cdot 1$$

In general, $\Gamma(t + 1) = t!$.

For this reason $\Gamma(t)$ is sometimes called the factorial function.

B) The Beta Function

The beta function, $\beta(t,1)$, is defined by

$$\beta(t,1) = \int_0^1 x^{t-1}(1 - x)^{1-1}\,dx \tag{9.9}$$

If $t > 0$, $1 > 0$ and either $t < 1$ or $y < 1$ (or both) the integral is improper but convergent.

If $t \geq 1$ and $1 \geq 1$, the integral is proper. An important relationship between the beta function and the gamma function is

$$\beta(t,1) = \frac{\Gamma(t)\,\Gamma(1)}{\Gamma(t + 1)} \tag{9.10}$$

9.10 STIRLING'S FORMULA

Stirling's formula states that if n is an integer, for large n, n! is approximately equal to

$$n! \sim \sqrt{2\pi n}\; n^n e^{-n} \tag{9.11}$$

where $\sim$ means "is approximately equal to".

CHAPTER 10

FOURIER SERIES AND INTEGRALS

10.1 PERIODIC FUNCTIONS

A function $f(x)$ is said to be periodic with period p if:

a) $f(x)$ is defined for all x in R

b) $f(x) = f(x + np)$ for all $x \in R$, where n is any integer.

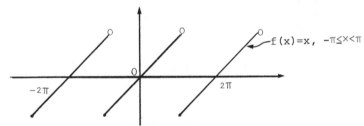

a. f periodic with period $p=2\pi$

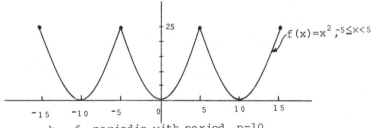

b. f periodic with period $p=10$

Fig. 10.1 Examples of periodic functions

98

The above definition states that functional values repeat themselves. This implies that the graph of a periodic function can be drawn for all x by repeating the graph on any interval p along the x-axis (Fig. 10.1 a,b).

10.2 FOURIER SERIES

If a function f is periodic with period p = 2a, then f may be represented by a form of infinite series, called the Fourier series:

$$f(x) = a_0 + \sum_{n=1}^{\infty} \left[a_n \cos\left(\frac{n\pi x}{a}\right) + b_n \sin\left(\frac{n\pi x}{a}\right) \right] \qquad (10.1)$$

where the Fourier coefficients a_0, a_n and b_n are

$$a_0 = \frac{1}{2a} \int_{-a}^{a} f(x)\,dx \qquad (10.2)$$

$$a_n = \frac{1}{a} \int_{-a}^{a} f(x)\cos\left(\frac{n\pi x}{a}\right) dx \qquad (10.3)$$

$$b_n = \frac{1}{a} \int_{-a}^{a} f(x)\sin\left(\frac{n\pi x}{a}\right) dx \qquad (10.4)$$

For the function $f(x) = x$ (see Figure 10.1a) with period 2π:

$$a_0 = \frac{1}{2\pi} \int_{-\pi}^{\pi} f(x)\,dx = \frac{1}{2\pi} \int_{-\pi}^{\pi} x\,dx = \frac{1}{2\pi} \left[\frac{\pi^2}{2} - \frac{\pi^2}{2}\right] = 0$$

$$a_n = \frac{1}{\pi} \int_{-\pi}^{\pi} f(x)\cos(nx)\,dx = \frac{1}{\pi} \int_{-\pi}^{\pi} x\cos(nx)\,dx = \frac{1}{\pi} \left[\frac{\cos(nx)}{n^2} + \frac{x\sin(nx)}{n}\right]_{-\pi}^{\pi} = 0$$

$$b_n = \frac{1}{\pi} \int_{-\pi}^{\pi} f(x)\sin(nx)\,dx = \frac{1}{\pi} \int_{-\pi}^{\pi} x\sin(nx)\,dx = \frac{1}{\pi} \left[\frac{\sin(nx)}{n^2} - \frac{x\cos(nx)}{n} \right]_{-\pi}^{\pi}$$

$$= \frac{1}{\pi} \frac{(-2\pi)\cos n}{n} = \frac{2}{n}(-1)^{n+1}$$

Thus using equation (10.1), the function $f(x) = x$ becomes

$$f(x) = \sum_{n=1}^{\infty} \frac{2(-1)^{n+1}}{n} \sin(nx) = 2\left(\sin x - \frac{1}{2}\sin(2x) + \frac{1}{3}\sin(3x) - \ldots \right)$$

Half Range Expansions

a) If the function $f(x)$ is periodic and even on the interval $(-a,a)$ then

$$f(x) = a_0 + \Sigma\, a_n \cos\left(\frac{n\pi x}{a} \right) \qquad (10.5)$$

where $a_0 = \frac{1}{a} \int_0^a f(x)\,dx$, $a_n = \frac{2}{a} \int_0^a f(x)\cos\left(\frac{n\pi x}{a} \right) dx$

With these values of the coefficients, the series (10.5) is known as the Fourier cosine series of $f(x)$ in $0 < x < a$.

If $f(x)$ is continuous and bounded on $0 < x < a$ and has a sectionally continuous derivative, and if $f(a+) = f(a-) = 0$, then the Fourier series converges uniformly to $f(x)$ in the interval $0 \le x \le a$.

b) If the function $f(x)$ is periodic and odd $f(-x) = -f(x)$, on the interval $-a < x < a$, and $f(0) = 0$, then

$$f(x) = \Sigma\, b_n \sin\left(\frac{n\pi x}{a} \right) dx \qquad (10.6)$$

where

$$b_n = \frac{2}{a} \int_0^a f(x)\sin\left(\frac{n\pi x}{a} \right) dx$$

The series (10.6) is known as the Fourier sine series of $f(x)$ in the interval $0 < x < a$.

If f is ;continuous and bounded on $0 < x < a$ and has a sectionally continuous derivative and if $f(a+) = f(a-) = 0$, then the equation (10.6) converges uniformly to f in the interval $0 \leq x \leq a$. The series converges to 0 at $x = 0$ and $x = a$.

10.3 COMPLEX NOTATION FOR FOURIER SERIES

as The Fourier series for the function $f(x)$ can be written

$$f(x) = \sum_{n=-\infty}^{\infty} c_n e^{in\pi x/a}, \qquad (10.7)$$

where

$$c_n = \frac{1}{2a} \int_{-a}^{a} f(x)e^{-in\pi x/a}dx$$

10.4 CONVERGENCE OF FOURIER SERIES

If $f(x)$ is piecewise continuous with period $p = 2a$, then the series $a_0 + \sum_{n=1}^{\infty} a_n \cos\left(\frac{n\pi x}{a}\right) + b_n \sin\left(\frac{n\pi x}{a}\right)$, with coefficients from the equations (10.2), (10.3) and (10.4), converges pointwise to $\frac{1}{2}\{f(x_0+) + f(x_0-)\}$ if x_0 is a point of discontinuity and $f(x_0)$ if x_0 is a point of continuity.

A) Uniform Convergence

Let $f(x)$ be continuous on an interval $a < x < b$ with period $p = 2a$. Suppose that $f'(x)$ is sectionally (same as piecewise) continuous on $a < x < b$. Then the uniform convergence theorem states that the series $a_0 + \sum_{n=1}^{\infty}\left[a_n \cos\left(\frac{n\pi x}{a}\right) + b_n \sin\left(\frac{n\pi x}{a}\right)\right]$ converges uniformly to f on $a < x < b$.

B) Mean Error and Convergence in the Mean

Let $f_n(x) = A_0 + \sum_1^N \left[A_n \cos(n\pi x/a) + B_n \sin(n\pi x/a) \right]$ be a sequence of functions whose limit is the Fourier series of $f(x)$. Let $f(x)$ be a function defined in the interval $-a < x < a$ for which $\int_{-a}^{a} f^2(x)dx$ is a finite number. Then:

a) Among all the sequence of the functions $f_n(x)$, the one which best approximates $f(x)$ in the sense of the error $\int_{-a}^{a} [f(x) - f_n(x)]^2 dx$ is the Fourier series

$$a_0 + \sum_1^N [a_n \cos(n\pi x/a) + b_n \sin(n\pi x/a)] \qquad (10.8)$$

b) $\dfrac{1}{a} \int_{-a}^{a} f^2(x)dx = 2a_0^2 + \sum_1^\infty [a_n^2 + b_n^2] \qquad (10.9)$

(Parseval's Identity)

$$a_n = \frac{1}{a} \int_{-a}^{a} f(x)\cos(n\pi x/a)dx \to 0$$

$$b_n = \frac{1}{a} \int_{-a}^{a} f(x)\sin(n\pi x/a)dx \to 0 \text{ as } n \to \infty$$

10.5 FOURIER INTEGRAL

A function $f(x)$ defined on R can often be written:

$$f(x) = \int_0^\infty (A(\alpha)\cos\alpha x + B(\alpha)\sin\alpha x)d\alpha \qquad (10.10)$$

The equality holds if $\int_{-\infty}^{\infty} |f(x)|dx$ exists and if in any finite interval there are only a finite number of discontinuities in f and/or f'. At discontinuities of f, this integral has value $\frac{f(x+0)+f(x-0)}{2}$, not necessarily $f(x)$.

(10.10) can be written:

$$f(x) = \frac{1}{\pi} \int\limits_{a=0}^{\infty} \int\limits_{u=-\infty}^{\infty} f(u)\cos a(x-u)\,du\,da \qquad (10.11)$$

or $\quad f(x) = \frac{1}{2\pi} \int\limits_{-\infty}^{\infty} \left[e^{-iax}da \int\limits_{-\infty}^{\infty} f(u)e^{iau}du \right] \qquad (10.12)$

If f is odd, we can simplify the above to

$$f_s(x) = \frac{2}{\pi} \int\limits_0^{\infty} \left[\sin ax\,da \int\limits_0^{\infty} f(u)\sin au\,du \right] \qquad (10.13)$$

Similarly if f is even,

$$f_c(x) = \frac{2}{\pi} \int\limits_0^{\infty} \left[\cos ax\,da \int\limits_0^{\infty} f(u)\cos au\,du \right] \qquad (10.14)$$

The inner integrals of (10.12), (10.13), and (10.14) are called Fourier Transforms, and have interesting properties. To make the Fourier Transform and inverse Fourier Transform symmetric, we use the square root of the outside constants in each.

10.6 FOURIER TRANSFORM

$$\tilde{f}(a) = \frac{1}{\sqrt{2\pi}} \int\limits_{-\infty}^{\infty} f(u)e^{iau}du \qquad (10.15)$$

Inverse Fourier Transform (the outer integral):

$$f(x) = \frac{1}{\sqrt{2\pi}} \int\limits_{-\infty}^{\infty} e^{-iax}\,\tilde{f}(a)da \qquad (10.16)$$

And Fourier sine and cosine transforms are formed by taking square roots of $\frac{2}{\pi}$ onto (10.13) and (10.14).

If $\tilde{f}(a)$ is the Fourier Transform of $f(x)$ and $\tilde{g}(a)$ the Fourier Transform of $g(x)$, then

$$\int_{-\infty}^{\infty} \tilde{f}(a)\tilde{g}(a)e^{-iax}da = \int_{-\infty}^{\infty} f(y)g(x-y)dy \qquad (10.17)$$

This is one version of the Convolution Theorem. Another version is:

$$F(f*g) = F(f)F(g) \qquad (10.18)$$

where $f*g$ means $\dfrac{1}{\sqrt{2\pi}} \displaystyle\int_{-\infty}^{\infty} f(y)g(x-y)dy$ and is called the convolution of f and g, and $F(h)$ means the Fourier Transform of $h(x)$.

Another interesting property of the Fourier Transform is the Parseval Identity:

$$\int_{-\infty}^{\infty} \tilde{f}(a)\overline{\tilde{g}(a)}da = \int_{-\infty}^{\infty} f(x)\overline{g(x)}dx \qquad (10.19)$$

and if $f = g$,

$$\int_{-\infty}^{\infty} |\tilde{f}(a)|^2 da = \int_{-\infty}^{\infty} |f(x)|^2 dx \qquad (10.20)$$

The Parseval Identity for sine and cosine Fourier Transforms is the same. Replace $\tilde{f}$ with $\tilde{f}_s$ or $\tilde{f}_c$.

Note: For f defined on semi-infinite intervals ($0 < x < \infty$), you may extend f to the region ($0 > x > -\infty$) by making f odd or even. Then you can use the sine or cosine series, whichever is convenient.

Note: The Laplace Transform is just the Fourier Transform (10.15) of a function which vanishes for $x < 0$. Replace (ia) with (s) real, and replace u with x, and multiply by $\sqrt{2\pi}$.

CHAPTER 11

FUNCTIONS OF A COMPLEX VARIABLE

11.1 COMPLEX FUNCTIONS

Let S be a set of complex numbers. A function f defined on S is a rule which assigns to each z in S a complex number w. The number w is called the value of f at z and is donated by f(z). The function f(z) can be written as

$$w = f(z) = u(x,y) + iv(x,y) \qquad (11.1)$$

where u and v are real valued functions of x and y.

The function f(z) is single-valued if for each value of z there corresponds only one value of w; otherwise it is multiple valued which can be considered as a collection of single valued functions.

Elementary Functions of a Complex Variable

a) The exponential function

The exponential function, e^z, is defined in the entire complex plane as

$$e^z = 1 + z + \frac{z^2}{2!} + \frac{z^3}{3!} + \ldots = e^x(\cos y + i\sin y) \quad (11.2)$$

In polar form equation (11.2) can be written as

$$e^z = \rho(\cos\phi + i\sin\phi) \quad \text{where} \quad \rho = e^x, \quad \phi = y \ (11.3)$$

b) Trigonometric Functions

$$\sin z = \frac{e^{iz} - e^{-iz}}{2i} = z - \frac{z^3}{3!} + \frac{z^5}{5!} -$$

$$\dots = \sum_{n=0}^{\infty} (-1)^n \frac{z^{n+1}}{(2n+1)!}$$
(11.4)

$$\cos z = \frac{e^{iz} + e^{-iz}}{2} = 1 - \frac{z^2}{2!} + \frac{z^4}{4!} - \dots$$

$$= \sum_{n=0}^{\infty} (-1)^n \frac{z^{2n}}{(2n)!}$$
(11.5)

The other four trigonometric functions are defined by the usual relations

$$\tan z = \frac{\sin z}{\cos z} , \qquad \sec z = \frac{1}{\cos z}$$

$$\cot z = \frac{\cos z}{\sin z} , \qquad \csc z = \frac{1}{\sin z}$$
(11.6)

From the definition of $\sin z$ it follows that

$$\sin z = \frac{e^{i(x+iy)} - e^{-i(x+iy)}}{2i}$$

$$= \sin x \left(\frac{e^y + e^{-y}}{2} \right) + i \cos x \left(\frac{e^y - e^{-y}}{2} \right)$$
(11.7)

Other properties of trigonometric functions are:

$$\sin^2 z + \cos^2 z = 1$$

$$\sin(z_1 + z_2) = \sin z_1 \cos z_2 + \cos z_1 \sin z_2$$

$$\cos(z_1 + z_2) = \cos z_1 \cos z_2 - \sin z_1 \sin z_2$$

$$\sin(-z) = -\sin z, \quad \cos(-z) = \cos z$$

$$\sin \left(\frac{\pi}{2} - z \right) = \cos z$$

$$\sin 2z = 2 \sin z \cos z$$

$$\cos 2z = \cos^2 z - \sin^2 z$$

If $z = n\pi$, $\sin z = 0$ $\qquad (n = 0, \pm 1, \pm 2, \dots)$

If $z = (n + \frac{1}{2})\pi$, $\cos z = 0$ $\qquad (n = 0, \pm 1, \pm 2, \dots)$

c) Hyperbolic Functions

$$\sinh z = \frac{e^z - e^{-z}}{2} = z + \frac{z^3}{3!} + \frac{z^5}{5!} + \dots$$

$$= \sum_{n=0}^{\infty} \frac{z^{2n+1}}{(2n+1)!}$$
(11.8)

$$\cosh z = \frac{e^z + e^{-z}}{2} = 1 + \frac{z^2}{2!} + \frac{z^4}{4!} + \dots$$

$$= \sum_{n=0}^{\infty} \frac{z^{2n}}{(2n)!}$$
(11.9)

$$\tanh z = \frac{\sinh z}{\cosh z} \qquad \operatorname{sech} z = \frac{1}{\cosh z}$$

$$\coth z = \frac{\cosh z}{\sinh z} \qquad \operatorname{csch} z = \frac{1}{\sinh z}$$
(11.10)

Further properties of hyperbolic functions are:

$$\cosh^2 z - \sinh^2 z = 1$$

$$\sinh(z_1 + z_2) = \sinh z_1 \cosh z_2 + \cosh z_1 \sinh z_2$$

$$\cosh(z_1 + z_2) = \cosh z_1 \cosh z_2 + \sinh z_1 \sinh z_2$$

$$\sinh(-z) = -\sinh z \qquad \cosh(-z) = \cosh z$$

$$\sinh(iz) = i\sin z \qquad \cosh(iz) = \cos z$$

$$\sin(iz) = i\sinh z \qquad \cos(iz) = \cosh z$$

d) The Logarithmic Function

$$\log z = \log r + i(\theta + 2n\pi), \text{ where} \qquad (11.11)$$

$$r = |z|$$

$$\theta = \arg z$$

$$n = 0, \pm 1, \pm 2, \pm 3, \dots$$

The principal value of logz is the value obtained from formula (11.11) when n = 0.

Further properties of logarithms are:

$$e^{\log z} = z$$

$$\log e^z = \log |e^z| + i \arg e^z = x + i(y + 2n\pi) = z + 2n\pi i$$

$$n = 0, \pm 1, \pm 2, \pm 3, \ldots$$

$$\log(z_1 z_2) = \log z_1 + \log z_2$$

$$\log \left[\frac{z_1}{z_2} \right] = \log z_1 - \log z_2$$

$$\log \left[z^{1/n} \right] = \frac{1}{n} \log z$$

e) The Power Function

$$f(z) = z^a = e^{a \log z}, \tag{11.12}$$

where a may be real or imaginary.

f) Inverse Trigonometric Functions

$$\sin^{-1} z = -i \log[iz + (1 - z^2)^{\frac{1}{2}}] \tag{11.13}$$

$$\cos^{-1} z = -i \log[z + i(1 - z^2)^{\frac{1}{2}}] \tag{11.14}$$

$$\tan^{-1} z = \frac{i}{2} \log \frac{i + z}{i - z} \tag{11.15}$$

$$\sinh^{-1} z = \log[z + (z^2 + 1)^{\frac{1}{2}}] \tag{11.16}$$

$$\cosh^{-1} z = \log[z + (z^2 - 1)^{\frac{1}{2}}] \tag{11.17}$$

$$\tanh^{-1} z = \frac{1}{2} \log \frac{1 + z}{1 - z} \tag{11.18}$$

11.2 LIMITS OF COMPLEX FUNCTIONS

Let a function $f(z)$ be defined at all points in some neighborhood of z_0, except possibly for the point z. itself.

The limit of f(z) as z approaches z_0 is defined as

$$\lim_{z \to z_0} f(z) = W_0 \qquad\qquad (11.19)$$

Statement (11.19) means that for each positive number ε there is a positive number δ such that

$$|f(z) - w_0| < \varepsilon \qquad \text{whenever} \quad 0 < |z - z_0| < \delta$$

Geometrically this definition means that for each ε in the neighborhood $|W - W_0| < \varepsilon$ of W_0 there is a δ neighborhood $|z - z_0| < \delta$ of z_0 such that the images of all points in the δ neighborhood, with the possible exception of z_0, lie in the ε neighborhood.

Theorem on Limits

a) If $f(z) = u(x,y) + iv(x,y)$, $z_0 = x_0 + iy_0$, $w_0 = u_0 + iv_0$

Then $$\lim_{z \to z_0} f(z) = w_0,$$

if and only if

$$\lim_{(x,y) \to (x_0,y_0)} u(x,y) = u_0 \quad \text{and} \quad \lim_{(x,y) \to (x_0,y_0)} v(x,y) = v_0$$

b) If $\lim_{z \to z_0} f(z) = w_0$ and $\lim_{z \to z_0} g(z) = w_0'$,

then

$$\lim_{z \to z_0} [f(z) + g(z)] = w_0 + w_0'$$

$$\lim_{z \to z_0} [f(z) \cdot g(z)] = w_0 w_0'$$

$$\lim_{z \to z_0} \frac{f(z)}{g(z)} = \frac{w_0}{w_0'} \quad \text{if } w_0' \neq 0$$

11.3 CONTINUITY

A function f(z) is continuous at a point z_0 if the following conditions are satisfied:

$$\lim_{z \to z_0} f(z) \quad \text{exists}$$

$$f(z_0) \quad \text{exists} \tag{11.20}$$

$$\lim_{z \to z_0} f(z) = f(z_0)$$

Statement (11.20) says that for each positive number ε there is a positive number δ such that

$$\left| f(z) - f(z_0) \right| < \varepsilon, \quad \text{whenever} \quad \left| z - z_0 \right| < \delta$$

The function, $f(z)$, is said to be continuous in a region R if it is continuous at each point in R.

11.4 CAUCHY-RIEMANN EQUATIONS

In order that $f(z) = u(x,y) + iv(x,y)$ be differentiable, or analytic in a region R of the complex plane, it is necessary that the Cauchy-Riemann equations

$$\frac{\partial u}{\partial x} = \frac{\partial v}{\partial y}, \tag{11.21}$$

$$\frac{\partial u}{\partial y} = -\frac{\partial v}{\partial x} \tag{11.22}$$

must be satisfied for values of x and y corresponding to all points in the region R.

From equations (11.21) and (11.22) the additional relations are obtained:

$$\frac{\partial^2 u}{\partial x^2} = \frac{\partial^2 v}{\partial x \partial y} \;,\; \frac{\partial^2 v}{\partial y^2} = \frac{\partial^2 u}{\partial y \partial x} \;,\; \frac{\partial^2 u}{\partial y^2} = -\frac{\partial^2 v}{\partial y \partial x} \;,\; \frac{\partial^2 v}{\partial x^2} = -\frac{\partial^2 u}{\partial x \partial y}$$

$$\tag{11.23}$$

If the second partial derivatives of u and v shown above are continuous, then

$$\nabla^2 u = \frac{\partial^2 u}{\partial x^2} + \frac{\partial^2 u}{\partial y^2} = 0 \tag{11.24}$$

110

$$\nabla^2 v = \frac{\partial^2 v}{\partial x^2} + \frac{\partial^2 v}{\partial y^2} = 0 \qquad (11.25)$$

Hence, it follows that the real and imaginary parts of an analytic function satisfy Laplace's equations. Functions satisfying Laplace's equation are called harmonic functions, and v is often called the harmonic conjugate of u.

11.5 ANALYTIC FUNCTIONS OF A COMPLEX VARIABLE

The derivative of a function of a complex variable is defined by the equation

$$\frac{df(z)}{dz} = f'(z) = \lim_{\Delta z \to 0} \frac{f(z + \Delta z) - f(z)}{\Delta z} \qquad (11.26)$$

when the limit exists.

A function $f(z)$ is said to be analytic in a region R of the complex plane if $f(z)$ has a finite derivative at each point of R and if $f(z)$ is single valued in R.

11.6 LINE INTEGRALS OF COMPLEX VARIABLES

If $f(z)$ is a single valued continuous function in a region R and C is a contour (a piecewise smooth curve) in the complex plane joining the points $z_1 = x_1 + iy_1$ and $z_2 = x_2 + iy_2$, the integral of $f(z)$ is defined as

$$\int_C f(z)dz = \int_C (u+iv)(dx+idy) = \int_C (udx-vdy) + i \int_C (vdx+udy)$$

$$(11.27)$$

A useful inequality of line integrals states that

$$\left| \int_C f(z)dz \right| \leq \int_C |f(z)| \ |dz| \leq M \int_C ds = ML \qquad (11.28)$$

where M is an upper bound for $|f(z)|$ along C and L is the length of the contour C.

If u and v are expressed in polar coordinates, equation (11.27) may be replaced by the equation

$$\int_C f(z)dz = \int_C [u\cos\theta - v\sin\theta)dr - (v\cos\theta + u\sin\theta)rd\theta]$$

$$(11.29)$$

$$+ i \int_C [(v\cos\theta + u\sin\theta)dr + (u\cos\theta - v\sin\theta)rd\theta]$$

A line integral over a closed curve is often denoted $\oint_C f(z)dz$.

11.7 CAUCHY'S THEOREM

Let C be a closed curve. If f(z) is analytic inside and on the curve C, then Cauchy's theorem states that

$$\oint_C f(z)dz = 0 \qquad (11.30)$$

11.8 CAUCHY'S INTEGRAL FORMULA

The value of f(z) at $z = \alpha$ in terms of the values of f(z) along an enclosing curve C can be written as

$$f(\alpha) = \frac{1}{2\pi i} \oint_C \frac{f(z)}{(z - \alpha)} \, dz \qquad (11.31)$$

As long as f is analytic within and on C, which is traversed counterclockwise. We can, in fact, find the nth derivative similarly:

$$f^{(n)}(\alpha) = \frac{n!}{2\pi i} \oint_C \frac{f(z)}{(z - \alpha)^{n+1}} \, dz$$

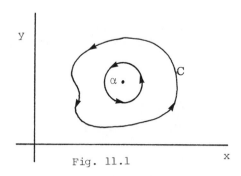

Fig. 11.1

11.9 TAYLOR SERIES

If $f(z)$ is analytic at a point $z = \alpha$, and on a circle around it, derivatives of $f(z)$ of all orders exist at that point, then $f(z)$ can be expanded by the Taylor series

$$f(z) = \sum_{n=0}^{\infty} \frac{f^{(n)}(a)}{n!} (z - a)^n = f(a) + f'(a)(z - a) + \frac{f''(a)}{2!}(z - a)^2$$

$$= \frac{f'''(a)}{3!} (z - a)^3 + \ldots \qquad (11.32)$$

which converges uniformly to f in the circle.

If the circle R of radius r is the largest circle around α such that $f(z)$ is analytic inside R, then r is called the radius of convergence of the Taylor series around α. Outside R the series diverges. On the boundary of R we don't know.

11.10 SINGULARITIES
OF ANALYTIC FUNCTIONS

Points at which a single valued function $f(z)$ is not analytic are called singular points or singularities of the function.

If $f(z)$ is analytic everywhere in some region except at an interior point $z = z_0$, we call z_0 an isolated singularity of $f(z)$.

11.11 LAURENT SERIES

Let R be a circular ring (or annulus) for which $\rho_1 \leq |z - z_0| \leq \rho_2$ such that $f(z)$ is analytic in R and on its inner and outer circular boundaries C_1 and C_2 (Fig. 11.2). Then at each point z in that domain, $f(z)$ is represented by the Laurent series.

$$f(z) = \sum_{n=0}^{\infty} a_n (z - z_0)^n + \sum_{n=1}^{\infty} \frac{b_n}{(z - z_0)^n} \qquad (11.33)$$

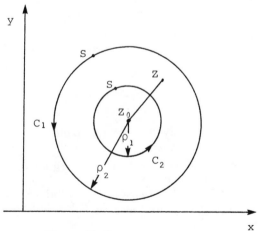

Fig. 11.2

where

$$a_n = \frac{1}{2\pi i} \int_{C_1} \frac{f(s)ds}{(s - z_0)^{n+1}} \quad (n = 0,1,2,\ldots),$$

$$b_n = \frac{1}{2\pi i} \int_{C_2} \frac{f(s)ds}{(s - z_0)^{-n+1}} \quad (n = 1,2,\ldots),$$

where each path of integration is being taken counterclockwise.

If $f(z)$ is analytic at all points inside and on C_1, the function $f(z)/(z - z_0)^{-n+1}$ is analytic inside and on C_2, therefore the coefficient b_n has a value equal to zero, and the Laurent series is reduced to a Taylor series.

If z_0 is an isolated singular point of $f(z)$, there is a positive number, r_1, such that the function is analytic at each point z for which

$$0 < |z - z_0| < r_1$$

The portion of the series involving negative powers of $z - z_0$ is called the principal part of f at z_0.

11.12 RESIDUES AND POLES

The Laurent series can be written as

(11.34)

$$f(z) = \sum_{n=0}^{\infty} a_n (z - z_0)^n + \frac{b_1}{(z - z_0)} + \frac{b_2}{(z - z_0)^2} + \cdots$$

The complex number

$$b_1 = \frac{1}{2\pi i} \int_C f(z)dz \qquad (11.35)$$

is called the residue of $f(z)$ at an isolated singular point $z = z_0$, and its value is denoted by $Res(z_0)$.

The closed contour, C, around z_0 is described in the positive sense such that $f(z)$ is analytic on C and interior to it except at the point z_0 itself.

If the principal part of the Laurent series,

$$\sum_{n=1}^{\infty} \frac{b_n}{(z-z_0)^n} \, ,$$

contains at least one non-zero term but the number of such terms is finite, there exists a positive number m such that $b_m \neq 0$ and $b_{m+1} = b_{m+2} = \ldots = 0$. In this case the isolated singular point z_0 is called a pole of order m. If m = 1, the pole is called a simple pole.

Then, the Laurent series can be written as

$$f(z) = \sum_{n=0}^{\infty} a_n (z - z_0)^n + \frac{b_1}{z-z_0} + \frac{b_2}{(z-z_0)^2} + \ldots + \frac{b_m}{(z-z_0)^m} \qquad (11.36)$$

Equation (11.35) can be written as

$$\int_C f(z)dz = 2\pi i \, \text{Res}(z_0) \qquad (11.37)$$

with

$$\text{Res}(z_0) = \lim_{z \to z_0} \frac{1}{(m-1)!} \left[\frac{d^{m-1}}{dz^{m-1}} \{ (z-z_0)^m f(z) \} \right] \qquad (11.38)$$

Another method for finding the residue of a function f(z) at a pole z_0 is by writing the function f(z) as a quotient

$$f(z) = \frac{p(z)}{q(z)} \qquad (11.39)$$

where p and q are both analytic at z_0 and $p(z_0) \neq 0$. If $q(z) = 0$ and $q'(z) \neq 0$, the residue of f at the simple pole z_0 is given by the formula

$$b_1 = \frac{p(z_0)}{q'(z_0)} \qquad (11.40)$$

11.13 CAUCHY'S RESIDUE THEOREM

Let f(z) be analytic in a simple closed contour, except for a finite number of poles $z_1, z_2, z_3, \ldots, z_n$ interior to C.

If $B_1, B_2, B_3, \ldots, B_n$ denote the residues of $f(z)$ at those points, then

$$\int_C f(z)dz = 2\pi i(B_1 + B_2 + B_3 + \ldots + B_n) \qquad (11.41)$$

where C is described in the positive sense (Fig. 11.3).

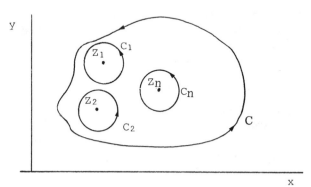

Fig. 11.3

11.14 EVALUATION OF REAL DEFINITE INTEGRALS

An important application of the theory of residues is the evaluation of certain types of real definite integrals.

We consider the most common types of improper integrals for which we can do this. These are:

A) Any real integral of the form

$$\int_0^{2\pi} R(\sin\theta, \cos\theta)d\theta \qquad (0 < \theta < 2\pi) \qquad (11.42)$$

where R is a rational function of $\sin\theta$ and $\cos\theta$, can be evaluated as follows:

117

a) Make the substitution

$$z = e^{i\theta} \qquad dz = ie^{i\theta} d\theta$$

there follows also

$$d\theta = \frac{dz}{iz}, \qquad \sin\theta = \frac{z^2 - 1}{2iz}, \qquad \cos\theta = \frac{z^2 + 1}{2z}$$

b) Then, the $R(\sin\theta,\cos\theta)d\theta$ takes the form $F(z)dz$, where C is the unit circle with center at origin, or

$$\int_C F(z)dz = 2\pi i \sum_k \text{Res}(z_k) \qquad (11.43)$$

where the points z_k are the poles of $F(z)$ inside the unit circle.

B) Any real integral of the form

$$\int_{-\infty}^{\infty} f(x)dx, \qquad (11.44)$$

where $f(x)$ is any real function. This integral can often be evaluated as follows:

a) Make the substitution $x = z$

b) Then consider the result of integrating $f(z)$ around the contour shown in Fig. 11.4.

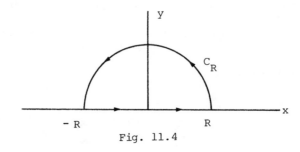

Fig. 11.4

For any value of R there follows

118

$$\int_{-R}^{R} f(x)dx + \int_{C_R} f(z)dz = 2\pi i \sum_{k} Res(z_k),$$

where the points z_k are the poles of $f(z)$ inside the contour.

c) Show that the integral taken along C_R tends to zero as $R \to \infty$

C) Any real integral of the form

$$\int_{-\infty}^{\infty} f(x)\cos mx\, dx \quad \text{or} \quad \int_{-\infty}^{\infty} f(x)\sin mx\, dx \qquad (11.45)$$

where $f(x)$ is a rational function. The integral can be evaluated as follows using the functions:

$$\cos mz = \frac{e^{imz} + e^{-imz}}{2} \quad \text{or} \quad \sin mz = \frac{e^{imz} - e^{-imz}}{2i},$$

the integral (11.45) takes the form

$$\int_{-\infty}^{\infty} e^{imx} f(x)dx = 2\pi i \sum_{k} Res\{e^{imz} f(z); z_k\} \quad (m \geq 0),$$

where the points z_k are the poles of $f(z)$ in the upper half plane (Fig. 11.4).

THE PROBLEM SOLVERS

The "PROBLEM SOLVERS" are comprehensive supplemental textbooks d signed to save time in finding solutions to problems. Each "PROBLEM SOLVER" is the fi of its kind ever produced in its field. It is the product of a massive effort to illustrate alm any imaginable problem in exceptional depth, detail, and clarity. Each problem is work out in detail with step-by-step solution, and the problems are arranged in order of complex from elementary to advanced. Each book is fully indexed for locating problems rapidly

ADVANCED CALCULUS
ALGEBRA & TRIGONOMETRY
AUTOMATIC CONTROL
   SYSTEMS/ROBOTICS
BIOLOGY
BUSINESS, MANAGEMENT,
  & FINANCE
CALCULUS
CHEMISTRY
COMPLEX VARIABLES
COMPUTER SCIENCE
DIFFERENTIAL EQUATIONS
ECONOMICS
ELECTRICAL MACHINES
ELECTRIC CIRCUITS
ELECTROMAGNETICS
ELECTRONIC COMMUNICATIONS
ELECTRONICS
FINITE & DISCRETE MATH
FLUID MECHANICS/DYNAMICS
GENETICS

GEOMETRY:
PLANE · SOLID · ANALYTIC
HEAT TRANSFER
LINEAR ALGEBRA
MACHINE DESIGN
MECHANICS : STATICS · DYNAMICS
NUMERICAL ANALYSIS
OPERATIONS RESEARCH
OPTICS
ORGANIC CHEMISTRY
PHYSICAL CHEMISTRY
PHYSICS
PRE-CALCULUS
PSYCHOLOGY
STATISTICS
STRENGTH OF MATERIALS &
  MECHANICS OF SOLIDS
TECHNICAL DESIGN GRAPHICS
THERMODYNAMICS
TRANSPORT PHENOMENA :
MOMENTUM · ENERGY · MASS
VECTOR ANALYSIS

If you would like more information about any of these books, complete the coup below and return it to us or go to your local bookstore.

RESEARCH and EDUCATION ASSOCIATION
61 Ethel Road W. • Piscataway • New Jersey 08854
Phone: (201) 819-8880

Please send me more information about your Problem Solver Books

Name _____

Address _____ _____

City _____ _____ State_____ Zip_____